Understanding the Elements of the Periodic Table™

CHROMIUM

Greg Roza

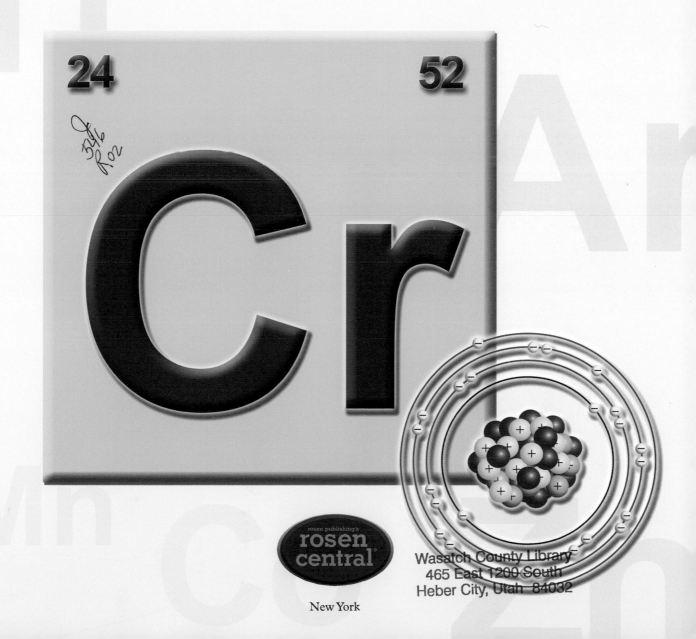

24

52

Cr

rosen publishing's
rosen
central®

New York

For Ira

Published in 2008 by The Rosen Publishing Group, Inc.
29 East 21st Street, New York, NY 10010

First Edition

Library of Congress Cataloging-in-Publication Data

Roza, Greg.
Chromium / Greg Roza.
      p. cm.—(Understanding the elements of the periodic table)
Includes bibliographical references and index.
ISBN-13: 978-1-4042-1411-8 (library binding)
1. Chromium. 2. Periodic law—Tables. I. Title.
QD181.C7R78 2008
546'.532—dc22

2007029652

Manufactured in Malaysia

On the cover: Chromium's square on the periodic table of elements; *(inset)* model of the subatomic structure of a chromium atom.

Contents

Introduction

In the 1920s, there was a race among several millionaires to build the world's tallest skyscraper. New York City, the fastest-growing city in the world at that time, became the center of this competition when Walter P. Chrysler decided to get involved. Chrysler was the founder of the successful Chrysler Corporation, a car manufacturing company. He believed that building the tallest skyscraper in the world would be a monument to him and to the car company he had built. He hired architect William Van Alen to complete the task. He soon discovered that Van Alen's ex-partner, H. Craig Severance, was attempting to build the tallest skyscraper, also in New York City, for a banker named George Ohrstrom.

Van Alen and Chrysler publicly announced that their skyscraper would be 925 feet (281.9 meters) tall. Severance and George Ohrstrom then planned for their building to be 927 feet (282.5 m) tall. However, Van Alan had a trick up his sleeve. Once the buildings were near completion, he revealed a 186-foot (56.6 m) spire that he had assembled inside the building! Once the spire was raised into place in 1930, the Chrysler Building reached a height of 1,046 feet (318.8 m), becoming the tallest building in the world.

The top of the Chrysler Building is made of stainless steel, which can't be manufactured without the element chromium. Chrysler and Van Alen chose stainless steel for its special qualities: it's shiny and silver, and the

This photo shows the top floors of the Chrysler Building. This gleaming edifice—a prime example of the Art Deco movement of the 1920s and 1930s—can be seen from many vantage points inside and outside the city of New York.

weather doesn't tarnish it. The building needed to be cleaned only twice in seventy-seven years. Chrysler wanted the top of the skyscraper to resemble the hubcaps, hood ornaments, and decorative features on the cars that were manufactured in his automobile plant. Many of these features also were made of stainless steel. The Chrysler Building held the tallest-building record for less than a year; it yielded that status to the Empire State Building in 1931. However, this gleaming landmark has remained a symbol of New York City and American capitalism.

This story emphasizes many of the characteristics of chromium that have made it one of the most important metals of the past 100 years. However, humans knew nothing about it until the mid-1700s. Even when chromium was discovered in the Ural Mountains, scientists did not really know what they had found.

Today, we can find chromium in many areas of our lives. Stainless steel is used to make countless products, such as cutlery, surgical tools, jewelry, and car parts, to name just a few. Other chromium products include paints and dyes, industrial chemicals, and a decorative and protective coating for other materials. In addition, the human body needs chromium to carry out several necessary biological functions. From the top of skyscrapers to the cells of our bodies, chromium is an element that we simply can't live without.

Chapter One
The History of Chromium

On July 26, 1761, German mineralogist Johann Gottlob Lehmann found an orange-red mineral in the Ural Mountains. He thought the mineral was a compound that contained lead, selenium, and iron. He named it Siberian red lead. However, it actually was lead chromate ($PbCrO_4$), which is called crocoite today. By the 1770s, Siberian red lead became a useful pigment in paints, but nobody realized there was an undiscovered element in it. It was used mainly to create red and yellow paints and fabrics.

Discovering Chromium

In 1797, French chemist Louis Nicolas Vauquelin studied Siberian red lead. He mixed the mineral with hydrochloric acid to create a dark green substance called chromium oxide (Cr_2O_3). He was able to isolate pure chromium metal by heating chromium oxide in an oven with charcoal. Further investigation revealed the presence of chromium in common gemstones. He learned that chromium gives emeralds their green color and rubies their red color. Vauquelin analyzed other minerals and discovered chromium in them. He named the new element chromium, from *chroma*, the Greek word for "color." He chose this name because chromium causes a variety of colors depending on the compound in which it is found.

In addition to being a chemist and professor of science, Louis Nicolas Vauquelin, shown here, also helped to write the pharmacy laws of his time.

For about 100 years, chromium was used strictly as a pigment in paints and clothes. One of the most popular chromium pigments was laboratory-prepared lead chromate. This synthetic compound commonly is called chromium yellow. Another chromium compound, potassium dichromate ($K_2Cr_2O_7$), was and still is used in the tanning of leather.

As chromium chemicals became more popular, mines and plants sprang up around the world. In Manchester, England, a chromium plant was created around 1808. After an important deposit of chromium ore was discovered on the border between Maryland and Pennsylvania in 1827, the United States became a major supplier of chromium for the next thirty years. Other important chromium sources were found in Turkey, India, and Africa.

The History of Stainless Steel

In 1821, French scientist Pierre Berthier found that when chromium was alloyed (mixed) with iron, the new metal could resist corrosion. However, it was too brittle to be of any use. During the next fifty years, other scientists experimented with combinations of iron, chromium, and other metals. In 1872, two Englishmen named Woods and Clark filed a patent for a

Louis Nicolas Vauquelin

Louis Nicolas Vauquelin (1763–1829) was a French chemist and pharmacist. In 1777, when he was just fourteen years old, Vauquelin became the apprentice of an apothecary (pharmacist). In 1783, he became the assistant of noted French chemist Antoine François, comte de Fourcroy. Vauquelin wrote many scientific papers with Fourcroy, although some did not bear his name. He began publishing his own scientific papers in 1790. In 1791, Vauquelin became a laboratory assistant at a well-known botanical garden in Paris called Jardin du Roi (now Jardin des Plantes). That same year, he became a member of the Academy of Sciences. He also helped to edit a chemistry journal.

During his life, Vauquelin held several respectable positions in the worlds of science and academics. As a professor, he lectured to many students who became renowned scientists in their own right. He discovered the elements beryllium and chromium. He was the first person to identify a number of basic chemicals. For example, he identified and isolated the first amino acid, asparagine, while studying asparagus. Vauquelin's contributions to science were honored when a genus of plants was named after him: Vauquelinia.

weather-resistant iron alloy that was 30 to 35 percent chromium and 2 percent tungsten. This alloy was one step closer to stainless steel, but it still lacked one essential ingredient.

In 1893, French chemist Henri Moissan heated ore that contained chromium and iron in an electric furnace with coke (carbon). The result was an alloy he called ferrochromium. This is a metal that contains up to 70 percent chromium and a small amount of carbon. The carbon helps to make the metal stronger, similar to the way it acts in steel. Ferrochromium would become the main ingredient in stainless steel, one of the most important metals of the twentieth century.

In the early 1900s, several scientists around the globe began experimenting with iron and chromium alloys. Some realized that in order to create an effective stainless steel, the carbon content had to be kept very low—around 0.15 percent. In 1911, German scientists P. Monnartz and W. Borchers discovered that 10.5 percent chromium was necessary for iron and chromium alloys to be effective in fighting corrosion. Their findings allowed other scientists to make great strides in developing stainless steel.

Many sources say that Harry Brearley, an English steelmaker, created the first stainless steel in 1913. The military asked Brearley to make gun barrels that would not rust. After experimenting with iron and chromium alloys, he developed a form of steel that contained 12.8 percent chromium and 0.24 percent carbon. However, other sources report that scientists in Germany, Poland, the United States, and Sweden created similar alloys between 1908 and 1913. No one is sure who created the first stainless steel.

Stainless Steel in the Twentieth Century

After 1913, the main use for chromium shifted from chemicals and pigments to the production of stainless steel. Production techniques improved, and scientists developed dozens of stainless steel grades. Nickel became a common additive to stainless steel; it increases the metal's ability to withstand corrosion. By 1920, stainless steel was being used to make a wide variety of everyday products, including jewelry, cookware, cutlery, and tools. It was used in the manufacture of car parts, particularly bumpers, radiators, and decorative trim. Other uses included coins, boat hulls, train bodies, and razor blades.

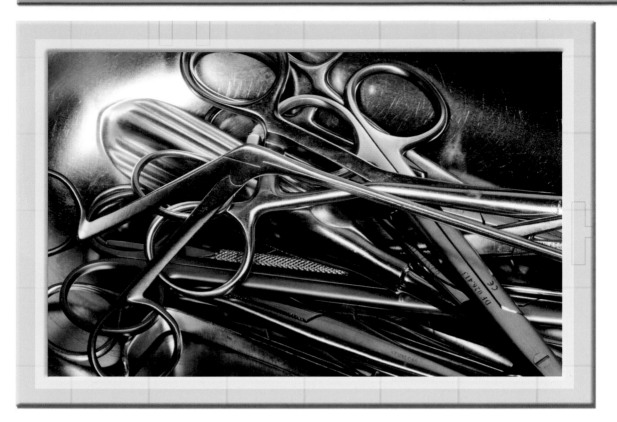

Many surgical tools, such as the forceps shown here, are made from a special type of stainless steel. Surgical stainless steel is an alloy of chromium, lead, carbon, molybdenum, and sometimes nickel. This alloy is easy to clean and sterilize, and it is highly durable.

Today, stainless steel is still used for most of these purposes, plus many more. Everywhere one looks one can see the lasting effect of the corrosion-resistant metal created in the early 1900s. Jet planes, spacecraft, nuclear plants, oil rigs, and countless skyscrapers could not have been built without stainless steel. Surgical tools, cookware, and cutlery are made of stainless steel because the metal makes them more durable and hygienic. Jewelry made of stainless steel is affordable and strong, and it lasts a long time.

Chromium in Our World

Stainless steel might be the most common use for chromium, but it isn't the only use. The many colors of chromium compounds have been important pigments in paints, dyes, inks, ceramics, glass, and building materials such as concrete. Chromium's ability to withstand corrosion has made it an important coating for other metals. Many industrial chemicals contain chromium. Our bodies even require chromium to function properly. Read on to find out more about chromium and its role in our lives.

Chapter Two
Chromium and the Periodic Table

All matter in the universe is made up of tiny building blocks called atoms. An atom is the smallest piece of matter that can exist by itself. Scientists have discovered 117 different kinds of atoms. Each kind of atom makes up one of the chemical elements, which are listed on the periodic table. Exploring the atom is similar to exploring the universe: it seems like there will always be some mystery surrounding it. However, scientists have made some amazing discoveries about atoms. These discoveries have changed the way we see the world and how we interact with it.

Subatomic Particles

Atoms are made up of even smaller parts called subatomic particles. There are three kinds of subatomic particles: protons, neutrons, and electrons. Protons and neutrons join to form a part of the atom called the nucleus. Protons have a positive electrical charge. Neutrons do not have an electrical charge. Most of the mass of an atom is in the nucleus.

Electrons are the smallest subatomic particles. They can be found in the space surrounding the nucleus. They orbit the nucleus in layers called shells. An atom can have between one and seven shells depending on the number of electrons orbiting the nucleus. A shell may contain a single electron, or it may contain dozens. The first shell—the one closest to the

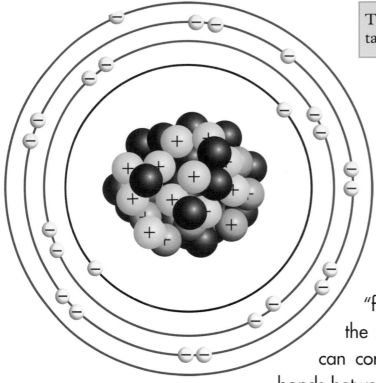

This diagram is a scientific representation of an atom of chromium.

nucleus—never has more than two electrons. As shells get farther from the nucleus, they can hold more and more electrons. For most atoms, the outermost shell is not full. Electron shells that are not "full"—meaning they don't contain the most possible electrons that they can contain—are capable of forming bonds between the atom and other atoms.

The most common chromium atom has twenty-four protons, twenty-eight neutrons, and twenty-four electrons. It has four electron shells. The first and second shells are full, holding two and eight electrons, respectively. The third shell has thirteen out of a maximum of eighteen electrons, and the outer shell has a single electron. The electrons in the outer two shells interact easily with other atoms, allowing chromium to form many different compounds with other elements.

What Is the Periodic Table?

The elements can be organized in a convenient arrangement called the periodic table. This table makes it easy for scientists and students to remember similarities and differences between chemical elements. The organization and number of subatomic particles determines what kind of matter their atoms make up and where they appear on the table. Most tables list just basic information for each of the elements.

By the mid-1800s, chemists needed a way to organize and remember the chemical elements that had been discovered. Several systems existed, but there was no unified method. In 1869, Russian chemist Dmitry Mendeleyev developed an early form of the periodic table that we are familiar with today.

Mendeleyev arranged the elements in order of increasing atomic weights (expressed as atomic mass units, or amu). This arrangement was such that certain traits occurred in regular intervals, or periods. Mendeleyev and the scientists who followed him left gaps in the periodic table where they believed other elements would belong once they were discovered. Many of these predictions proved to be true. Scientists are still working to identify undiscovered chemical elements, although new elements will go only at the end of the periodic table. All of the gaps in Mendeleyev's table have been filled.

Categorizing Chemical Elements

The periodic table lists elements in order of their atomic numbers. The atomic number is the number of protons in an atom of each element. The first element on the list is hydrogen. An atom of hydrogen has one proton, so its atomic number is 1. At the other end of the table, the element ununoctium has 118 protons, so its atomic number is 118. Chromium has twenty-four protons. Its atomic number is 24.

Atoms sometimes are categorized by their atomic weights. Atomic weight is a measure of an atom's mass. (Atomic weight also is called atomic mass.) An atom's atomic weight is the total weight of its protons, neutrons, and electrons. The atomic weight of an atom can be approximated by adding the weights of just the protons and neutrons because electrons are extremely light. Since chromium usually has twenty-four protons and twenty-eight neutrons, its atomic weight is often given as 52. However, chromium's atomic weight usually is listed as 51.9961 amu on

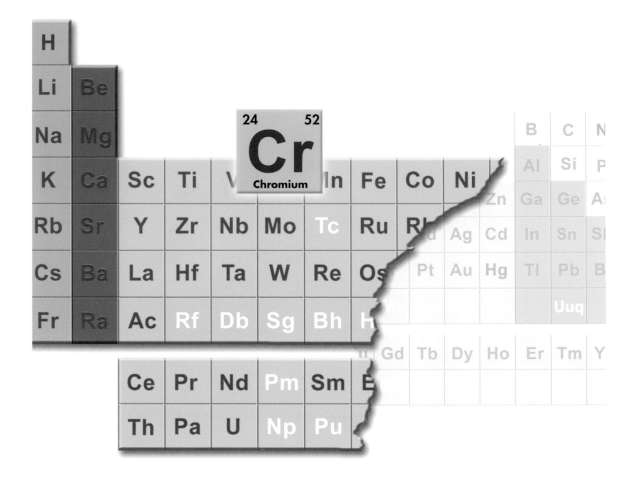

This section of the periodic table of elements highlights chromium's place on the table. Chromium is a transition metal located between vanadium and manganese.

the periodic table. This number is an average of the atomic weights of all the chromium isotopes. (See sidebar about isotopes.)

Groups, Periods, and Blocks

The elements are listed in the periodic table in rows and columns. Rows are called periods, and columns are called groups. There are eighteen groups in the periodic table. Chromium is in group VI, as are molybdenum and

What Is an Isotope?

Atoms of a particular element always have the same number of protons, but they can have different numbers of neutrons. Atoms with differing numbers of neutrons are called isotopes. An atom's atomic mass on the periodic table is an average of the atomic masses of all isotopes of the element. Some isotopes are stable, meaning they do not deteriorate into other chemical elements as time passes. Others are radioactive and sometimes dangerous to human beings. Many isotopes, both radioactive and stable, are helpful. The word "isotope" is Greek for "at the same place," since all isotopes of an element occupy the same place on the periodic table.

Chromium has twenty-eight isotopes. Normally, the element has twenty-eight neutrons. However, it can have as many as forty-three, or as few as eighteen. Natural chromium exists as one of four stable isotopes: 50Cr, 52Cr, 53Cr, and 54Cr. The numbers in these symbols stand for the total number of protons and neutrons in the nucleus.

All other chromium isotopes are radioactive. Radioactive isotopes release energy in the form of subatomic particles and high-energy radiation. This can happen slowly or quickly, depending on the isotope. The most stable of the radioactive chromium isotopes is 51Cr. It has a half-life of 27.7 days. All other chromium radioisotopes have a half-life of less than twenty-four hours. Some chromium radioisotopes have important medical applications.

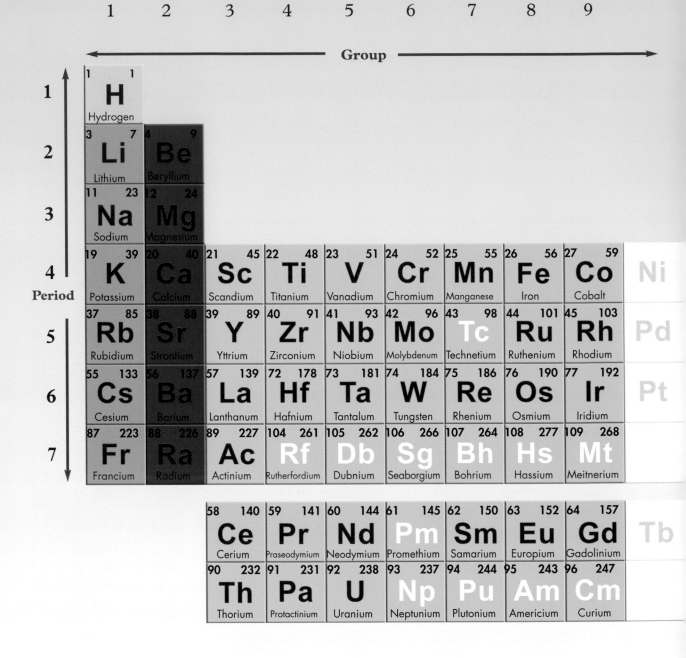

This periodic table shows the positions of the groups and periods. The yellow element (hydrogen) is a gas. The orange elements are alkali metals, and the red elements are alkaline earth metals. The green elements are transition metals. The elements that appear in white are man-made.

tungsten. All of these elements are silvery, shiny metals. They are known to have a role in the biological chemistry of living organisms. Seaborgium, a synthetic element, is in this group as well.

There are seven rows, or periods, in the periodic table. All elements in the same row have the same number of electron shells. Chromium is in the fourth period. All elements in period IV have four electron shells.

The Characteristics of Chromium

Pure chromium is hard, brittle, and bluish-silver. It has relatively high melting and boiling points. It conducts heat and electricity well. The presence of chromium in compounds causes a great variety of colors.

Chromium's outer two shells are not filled with electrons. Like the other transition metals, these incomplete shells allow chromium to play an important role in many chemical reactions. In fact, it often forms several different combinations with the same element.

Pure chromium does not react with water and oxygen in the air, which makes it resistant to weathering. This is why chromium is used as a coating to protect other metals from corrosion. When used for this purpose, chromium is called chrome. Automobiles that were manufactured in the 1950s used a lot of chrome. When polished, chrome becomes very shiny. It gives other metals a protective yet attractive covering.

Chapter Three

Chromium: From the Earth to the Lab

Chromium oxide is the ninth most abundant compound in the earth's crust. Chromium is the thirteenth most abundant element. All chromium ores are made up of igneous rocks—rocks that were once liquid. Chromium is also abundant in the earth's core. After millions of years, this liquid chromium oozed up through cracks in the earth's surface. As these pockets of chromium cooled, they formed deposits of solid chromium and chromium compounds. Chromium deposits occur in two shapes. The most common are thin layers called strata. Chromium may also occur in scattered blobs.

The largest chromium mine in the world is in South Africa. Other significant mines are located in Kazakhstan, India, Brazil, Finland, Turkey, and Zimbabwe. Some mines are beneath the earth's surface; others are on the surface. The only chromium ore is chromite, which is a mixed oxide of iron, magnesium, and chromium. Scientists believe there are about 11 billion tons of chromite waiting to be mined. This is enough to last for centuries. But pure chromium is rare in nature. The largest source of pure chromium is the Udachnaya diamond mine in Russia. The same types of forces that form diamonds also form pure chromium.

Making pure chromium metal is more time consuming and expensive than making ferrochromium. Partly because of this, it is not used as often as chromium alloys made with ferrochromium. There are several ways to

Chromium Snapshot

Chemical Symbol:	Cr
Classification:	Transition metal
Properties:	Steel-gray, lustrous, hard
Discovered by:	Louis Nicolas Vauquelin, 1797
Atomic Number:	24
Atomic Weight:	51.9961 atomic mass units (amu)
Protons:	24
Electrons:	24
Neutrons:	26, 28, 29, and 30
State of Matter at 68° Fahrenheit (20° Celsius):	Solid
Melting Point:	3,465°F (1,907°C)
Boiling Point:	4,840°F (2,671°C)
Commonly Found:	In the minerals chromite and crocoite

The mineral on the left is a chunk of chromium ore. The photo on the right shows several pieces of pure chromium. The luster of the pure chromium demonstrates why it is commonly used in the manufacture of chrome plating.

What Is Ferrochromium?

Most chromite ore is used to make a substance called ferro-chromium. This is an alloy of iron and chromium. Ferrochromium is an important alloy because it is used to make stainless steel.

To make ferrochromium, the chromite ore is first crushed into a powder. This makes it possible to remove most of the impurities that are in the ore when it is mined. The chromite powder is heated in a furnace with coke, or pure carbon fuel. This refining process is known as smelting. When it reaches around 3,600°F (2,000°C), the chromite melts and forms two liquids: iron oxide and chromium oxide. The carbon in the coke mixes with the oxygen in these two liquids to form a gas called carbon monoxide. This gas burns within the furnace, making the process even hotter and quicker. The chromium and iron mix to form ferrochromium.

make pure chromium. The most common method is known as the aluminothermic process. "Alumino" stands for the element aluminum, which plays an important role in the process, and "thermic" refers to a reaction that makes or absorbs a lot of heat.

First, chromite ore is mixed with sodium oxide to make sodium chromate (Na_2CrO_4), which is separated from the mixture by dissolving it in water. The solution is acidified and evaporated, producing solid sodium dichromate ($Na_2Cr_2O_7$). This is converted to chromium oxide (Cr_2O_3) by heating with sulfur. The chromium oxide is mixed with aluminum powder and ignited. This reaction creates an intense fire. When it reaches about 3,600°F (2,000°C), the oxygen atoms in the chromium oxide bond with the aluminum atoms, resulting in aluminum oxide (Al_2O_3) and pure chromium. Sometimes the aluminum is replaced with silicon in this process.

Pure chromium also can be refined using the electrolyte method. One of the most common ways of doing this is to dissolve chromium oxide in sulfuric acid. The solution that results is called an electrolyte. This is a solution that contains free ions—negatively charged atoms or groups of atoms—and is capable of conducting electricity. When two metal rods are placed in the solution and an electric current is passed between them, pure chromium builds up on the negatively charged rod.

Oxides and Chromates

All this talk about chromite and chromate might be confusing. As we learned, chromite is an ore made of iron, magnesium, and chromium. A chromate, on the other hand, is a type of chromium compound comprised of oxygen and other metallic elements. It is important to know that chromium comes in several main forms, depending on what it is mixed with. The element is capable of creating different compounds based on the number of electrons it loses or gains when forming bonds with other elements. Many common chromium compounds are often sorted into two main categories: oxides and chromates.

Chromium Oxides

Oxides are chemical compounds containing oxygen and one or more elements. Many oxides form when an element is exposed to air. Chromium oxides are hard crystals that form when chromium is exposed to oxygen. The bonds that form in chromium oxides are very strong. They are called ionic bonds. This means that the atom of chromium gives electrons to the atoms it is bonding to. An atom that loses or gains electrons—giving it a

This is an open-pit chromium mine in the Yamalo-Nenets Autonomous District of Russia. This sparsely populated area on the Arctic Circle is well known for its natural resources, especially oil and natural gas.

negative or positive electrical charge—is called an ion. Since electrons have a negative charge, the loss of electrons gives the chromium atom a positive charge. This type of ion is called a cation. The other atom gains electrons and has a negative electrical charge. This type of ion is called an anion. Negative and positive charges attract each other. This makes the bonds between cations and anions strong. Ionic bonds usually occur between metals and nonmetals.

Atoms don't always lose the same number of electrons when forming ionic bonds. When chromium bonds with oxygen to create chromium oxide, for example, it can lose two, three, or six electrons. The different numbers of electrons that an atom loses (or gains) in forming an ion produce different oxidation states. Chromium has three common oxidation states. These are indicated using Roman numerals. For example, there are three forms of chromium oxide based on chromium's oxidation states: chromium(II) oxide, chromium(III) oxide, and chromium(VI) oxide. Each of these forms has unique characteristics.

Chromates and Dichromates

Chromates are compounds that contain chromium, oxygen, and at least one other metal. Each chromium atom in a chromate bonds with four oxygen atoms to create a chromate ion. The chromium atom must share two electrons with each atom of oxygen (for a total of eight shared electrons). This type of bond is called a covalent bond.

Chromium has only six electrons to share: five electrons in its third shell and one electron in its fourth shell. So, the chromium atom needs to "borrow" two electrons from another atom. These borrowed electrons give the chromate ion a 2− charge, making it a complex anion (negatively charged ion). The chemical symbol for the chromate ion is $[CrO_4]^{2-}$.

The chromate ion forms ionic bonds with positive ions from other elements to form compounds. In the compound lead chromate (PbCrO4), for example, each lead atom provides two electrons for each CrO_4

This group of potassium dichromate crystals was made in a lab in the early 1920s.

complex. This makes two ions: a lead cation (Pb2+) and a chromate anion. These ions form ionic bonds between them. Like most chromates, lead chromate varies in color from bright yellow to orange-red.

Dichromate compounds contain an ion formed from two chromium atoms and seven oxygen atoms. These atoms form the dichromate anion $[Cr_2O_7]^{2-}$. Dichromate ions form ionic bonds with metals the same way that chromate ions do. For example, two potassium atoms each provide one electron to a dichromate ion to form potassium dichromate ($K_2Cr_2O_7$). Like most dichromates, potassium dichromate is a bright-orange color. Chromates and dichromates often are used in paints and dyes because of their bright colors.

Chapter Four
Chromium Compounds and Alloys

Pure chromium is somewhat brittle. For this reason, few useful objects are made of pure chromium. However, chromium compounds and alloys are very useful. In the previous chapter, we learned some basic chemical facts about chromium. In this chapter, we will learn about common chromium compounds and alloys and how they are used in everyday life. For instance, chromium can be used as an agent in tanning and plating, which is a process for surfacing metal.

Chromite

Chromite is a compound of chromium oxide and either iron ($FeCr_2O_4$) or magnesium ($MgCr_2O_4$). It is the only important ore of chromium. Almost half of the chromite mined in the world is used in the production of stainless steel. Chromite is used to manufacture pure chromium and all the other chromium compounds in use today. It usually ranges from dark brown to black in color.

Chromite is a widely used refractory material. "Refractory" means that it is able to retain its strength at very high temperatures. Chromite can be used, for example, to manufacture bricks that line the walls of industrial furnaces.

Chromium(III) oxide, shown here, was first prepared by French scientists in 1838.

Chromium(III) Oxide

Chromium(III) oxide is made up of chromium ions (in the 3+ oxidation state) and oxide ions. This is the most stable and common chromium oxide. Chromium(III) oxide powder is manufactured from the mineral chromite. A common name for chromium oxide is chromia. Because of its emerald color, it is also called chrome green.

Chrome green is widely used as a coloring agent in many products. These include enamel, glass, leather, plastics, bar soaps, and concrete. It is valued for its vibrant color and its ability to withstand weathering and high temperatures. Chromium(III) oxide is also used as a catalyst in chemical reactions.

Other forms of chromium oxide are no longer as useful. Chromium dioxide, or chromium(IV) oxide, is a synthetic magnetic substance. It is used in the manufacture of magnetic tape cassettes for recording and storing sound and data. With the increasing popularity of CDs and DVDs, however, it has been used less and less for that purpose.

Crocoite

Chromium was first discovered in the mineral crocoite, which is mostly lead chromate ($PbCrO_4$). Most natural crocoite ore is found in Australia. Natural

This is a photograph of natural crocoite crystals. At one time, gem collectors flocked to mines in Australia and New Zealand for large crocoite crystals. Today, however, very few crocoite crystals are longer than two inches (about five centimeters).

crocoite crystals are long and thin, and have a bright, orange-red color. The crystals are prized by mineral collectors for their beauty and dazzling color.

Laboratory-prepared lead chromate is an orange-yellow powder. It is made by mixing a lead(II) salt with potassium chromate. Commonly called chrome yellow, it has been used as a pigment in paints and fabrics since the early nineteenth century.

Potassium Dichromate

Potassium dichromate ($K_2Cr_2O_7$) is an artificial compound. It has a deep, reddish-orange color. Like many chromium compounds, it is sometimes used

What Is an Alloy?

An alloy is a mixture of two or more elements, at least one of which is a metal. The combination is helpful because the best traits of the different elements work together to make a more effective substance. Alloys are not compounds. They are metals blended with other metals or nonmetals. The atoms in the mixture do not bond with each other as they do in compounds. Rather, they simply exist together as a mixture.

By mixing different elements together, substances can be produced that have different properties than the separate elements have on their own. Alloys are often stronger than regular metals. The atoms in a pure metal are spaced uniformly and are all the same size. This allows them to move easily over each other, so a pure metal is relatively easy to bend. The atoms in an alloy are different sizes and are spaced out randomly. This prevents the atoms from easily moving past each other. Atoms are squeezed up against each other in an alloy, which helps keeps them in place. This quality makes an alloy difficult to bend and break.

Bronze is one of the oldest alloys, known and used since prehistoric times. Brass is usually an alloy of copper and tin, and its bright-yellow color resembles gold. Other common alloys include steel (lead iron and carbon), white gold (gold and silver), sterling silver (silver and copper), and pewter (tin, lead, and copper). As you have learned, chromium is an ingredient in one of the most important alloys of the past 100 years—stainless steel.

as a pigment in paints and fabric dyes. It is also used as an oxidizing agent in chemical reactions. An oxidizing agent is a substance that changes another substance by taking electrons from it. Potassium dichromate is sometimes used as a catalyst when producing pure chromium metal. It can be found in glassware cleaners, photography chemicals, and fireworks. Potassium dichromate is a dangerous chemical that can irritate skin and eyes and can cause burns. It is believed to be a carcinogen.

Stainless Steel

As a metal, chromium is most valuable when it is alloyed with other metals. The most common chromium alloy is stainless steel, an alloy of iron, a small amount of carbon, and at least 10.5 percent chromium by weight. The carbon makes the steel harder, and the chromium makes the steel resistant to corrosion (rust). Normally, iron reacts with oxygen in damp air to form rust. As iron forms rust, the rust flakes off, exposing more iron to damp air. In the open air, an entire piece of iron will eventually turn to rust. The chromium in stainless steel also reacts with oxygen in the air. However, instead of flaking off, the chromium oxide forms a very thin coating on the surface that protects the object from further reaction with oxygen. The layer of chromium oxide can even "heal" itself when it becomes damaged, forming a renewed oxide coating. While chromium helps extend the life of stainless steel, the alloy will in fact turn to rust slowly, usually over a period of about 100 years.

Stainless steel's unique characteristics make it useful for many applications. It is commonly used for objects that must have hygienic surfaces—such as surgical tools and cookware—because its does not have cracks or pores that dirt and bacteria can adhere to. Stainless steel with a high content of chromium is able to resist damage by heat and fire. It also reflects light well, providing a shiny appearance to objects. This makes it a popular material for use in making attractive, inexpensive, and

The Walt Disney Concert Hall, designed by architect Frank Gehry, is covered in stainless steel plates. When the building was completed, the sun's glare off the highly polished stainless steel created "hot spots" on the Los Angeles sidewalks around the building—some which reached about 140°F (60°C). The surfaces were sanded to reduce the glare.

durable jewelry. Buildings, monuments, and sculptures made of stainless steel remain shiny and new looking despite being subjected to the elements.

There are more than 150 grades of stainless steel, but only about 15 are commonly used. Other elements sometimes used in the production of stainless steel include nickel, copper, titanium, and molybdenum. Stainless steel is recyclable. Most objects made of stainless steel today contain more than 50 percent recycled stainless steel.

As we have seen, chromium can be found in many places and in many forms. It is an essential ingredient in stainless steel, and its colorful compounds are used in pigments for paints, dyes, glass, and other applications. Pure chromium is commonly used to give other metals an attractive yet protective coating. Without chromium, our bodies simply couldn't function.

Chrome Plating

One of the most common uses for pure chromium is as a protective coating for other metals. Chromium applied in this manner is often called chrome. Chrome has been used for more than 100 years. You may have seen chrome on the fenders and decorative trim of automobiles.

The process used to coat objects in chrome is called electroplating. Usually, a metal object is immersed in an acid that contains chromium(VI) (meaning the atoms are missing a total of six electrons, giving them a positive electrical charge). The metal object and a chromium rod are placed in the acid. Then an electrical current, which is a flow of electrons, is passed though the acid bath. The electrons flow through the metal object, into the acid, and out of the chromium rod. Electrons in the electric current join with the chromium(VI), turning it into pure chromium metal.

Chrome bumpers, like those on this Cadillac Eldorado, were a popular feature on automobiles in the 1950s.

This metal forms a thin layer on the metal object.

This process is dangerous for several reasons. The acid and electricity can cause harm if they are not used carefully. Chromium(VI), as we will discuss later in this chapter, is very dangerous to human beings. However, once the process is complete, the metal object has a coating of chrome that will keep it from rusting. It can be polished to shine brightly.

Chromium and the Human Body

Chromium is a trace metal that occurs in the cells of plants and animals in very small quantities. Chromium plays a very important role in the body's ability to metabolize fats and carbohydrates and in creating energy from blood sugar. It may improve cholesterol levels, help build muscle, decrease fat, and help the body to use carbohydrates more efficiently.

Chromium also helps the hormone insulin do its job. The body releases insulin into the bloodstream after we eat. The insulin tells the tissues of the body to absorb glucose (sugar) from the blood and store it for future use. Chromium helps insulin move glucose from the bloodstream to the cells where it is needed. A chromium deficiency can cause the body to have difficulty absorbing glucose from the bloodstream. Because people with diabetes lack insulin, chromium may help prevent or cure type 2 diabetes. No one is sure just how much chromium the human body needs on a daily basis. The National Academy of Sciences recommends between 50 and

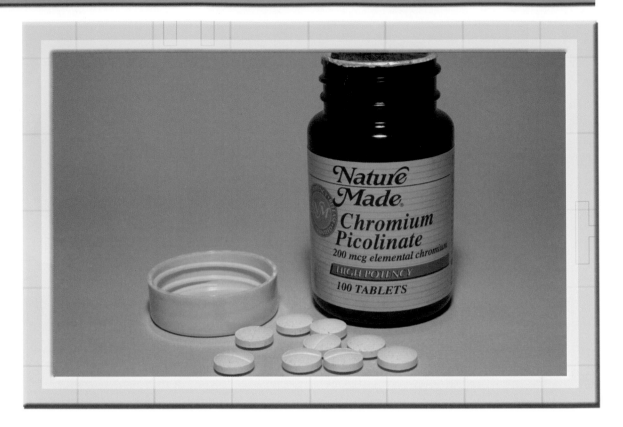

The manufacturers that make chromium picolinate supplements claim that it can reduce fat, build muscle, and help diabetics stay healthy. Some scientists, however, fear that it has not been studied well enough to be considered safe.

200 micrograms of chromium daily. Some suggest 500 is a better estimate. However, it would be very difficult to get that much chromium from food in one day. Because of this, some people prefer to take chromium supplements. Chromium picolinate is one of the most popular supplements on the market today.

Hexavalent Chromium

Hexavalent chromium compounds are those that contain chromium in the +6 oxidation state. In other words, each chromium atom in the compound has given up six electrons to other atoms. Although hexavalant chromium

Foods That Contain Chromium

Nutritionists recommend that adults get at least 120 micrograms of chromium in their diet every day. However, most people get only 35 micrograms. The best food source of chromium is brewer's yeast. The list below shows several other good sources of chromium.

Food	Serving Size	Chromium in Micrograms
onion	1 cup	24.8
broccoli	1 cup	22.0
tomato	1 cup	9.0
romaine lettuce	1 cup	7.9
grape juice	1 cup	7.5
potato	1 cup	2.7
green beans	1 cup	2.2
beef	3 ounces	2.0

Many body builders eat broccoli because the high amount of chromium in it helps build muscle and reduce fat.

compounds have many uses in modern society, they are known carcinogens when inhaled.

Hexavalent chromium compounds are used as pigments in paints, dyes, inks, plastics, photography, and fireworks. They are used in the production of stainless steel and leather. They also play an important role in the process of chrome plating. When chromium(VI) enters the tissues of the body (especially the lungs), it can lead to life-threatening health problems. People who are overly sensitive to hexavalent chromium may experience skin rashes when they come into contact with it. Inhaling hexavalent chromium can cause asthma and nosebleeds. Long-term exposure to the chemical can cause the breakdown of the nasal septum and can lead to lung cancer.

People who work around hexavalent chromium must wear appropriate protective equipment, particularly respirators. However, health standards were not always as strict as they are today. Improper handling of hexavalent chromium has resulted in many illnesses and deaths. When industrial plants using it are careless with its disposal, it can pollute water and air, leading to widespread illness.

Hexavalent chromium became a public story in the 1990s when Erin Brockovich, a law clerk from California, helped to establish a lawsuit against Pacific Gas and Electric Company (PG&E). The soil around one of PG&E's natural gas pipelines was found to be contaminated with hexavalant chromium, as was the drinking water for the nearby town of Hinkley. Many people in Hinkley during the 1960s, 1970s, and 1980s developed cancer. Brockovich and the

Ruth Ann Vaughn holds pictures of family members she believe died due to exposure to hexavalent chromium.

people of Hinkley contended that the contaminated water was the culprit. In the end, PG&E was ordered to pay 600 Hinkley residents a total of $333 billion dollars—the largest settlement ever paid out in a direct-action lawsuit in U.S. history.

Some scientists say there is no evidence to prove that hexavalent chromium can cause cancer when ingested (rather than inhaled). Others say it has caused cancer in laboratory studies. Either way, the highly publicized trial became the basis for the Academy Award–winning movie *Erin Brockovich*, which starred Julia Roberts. Thanks to Brockovich and the movie named after her, the world became aware of the dangers of hexavalent chromium.

Chromium in Your World

Even though hexavalent chromium is a dangerous chemical, most chromium compounds are perfectly safe when used properly. We use many objects made of chromium every day without even thinking about it. We prepare and eat our food with pans and cutlery that contain chromium. We wash ourselves with soap colored with chromium compounds. We wear chromium jewelry on our fingers and around our necks. Doctors use chromium tools and medicines to make us well. We even ingest chromium daily. Without chromium, our world would be a more laborious, less healthy, and certainly less colorful place.

The Periodic Table of Elements

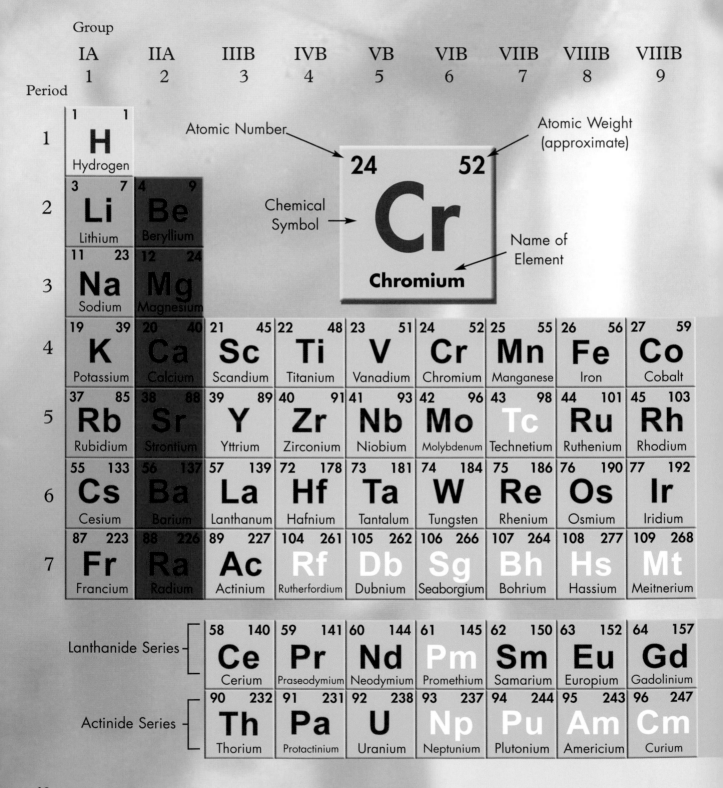

Group

IA	IIA	IIIB	IVB	VB	VIB	VIIB	VIIIB	VIIIB
1	2	3	4	5	6	7	8	9

Period

Atomic Number

Atomic Weight (approximate)

24		52
	Cr	
	Chromium	

Chemical Symbol

Name of Element

Period 1

1	1
H	
Hydrogen	

Period 2

3	7	4	9
Li		**Be**	
Lithium		Beryllium	

Period 3

11	23	12	24
Na		**Mg**	
Sodium		Magnesium	

Period 4

19 39	20 40	21 45	22 48	23 51	24 52	25 55	26 56	27 59
K Potassium	**Ca** Calcium	**Sc** Scandium	**Ti** Titanium	**V** Vanadium	**Cr** Chromium	**Mn** Manganese	**Fe** Iron	**Co** Cobalt

Period 5

37 85	38 88	39 89	40 91	41 93	42 96	43 98	44 101	45 103
Rb Rubidium	**Sr** Strontium	**Y** Yttrium	**Zr** Zirconium	**Nb** Niobium	**Mo** Molybdenum	**Tc** Technetium	**Ru** Ruthenium	**Rh** Rhodium

Period 6

55 133	56 137	57 139	72 178	73 181	74 184	75 186	76 190	77 192
Cs Cesium	**Ba** Barium	**La** Lanthanum	**Hf** Hafnium	**Ta** Tantalum	**W** Tungsten	**Re** Rhenium	**Os** Osmium	**Ir** Iridium

Period 7

87 223	88 226	89 227	104 261	105 262	106 266	107 264	108 277	109 268
Fr Francium	**Ra** Radium	**Ac** Actinium	**Rf** Rutherfordium	**Db** Dubnium	**Sg** Seaborgium	**Bh** Bohrium	**Hs** Hassium	**Mt** Meitnerium

Lanthanide Series

58 140	59 141	60 144	61 145	62 150	63 152	64 157
Ce Cerium	**Pr** Praseodymium	**Nd** Neodymium	**Pm** Promethium	**Sm** Samarium	**Eu** Europium	**Gd** Gadolinium

Actinide Series

90 232	91 231	92 238	93 237	94 244	95 243	96 247
Th Thorium	**Pa** Protactinium	**U** Uranium	**Np** Neptunium	**Pu** Plutonium	**Am** Americium	**Cm** Curium

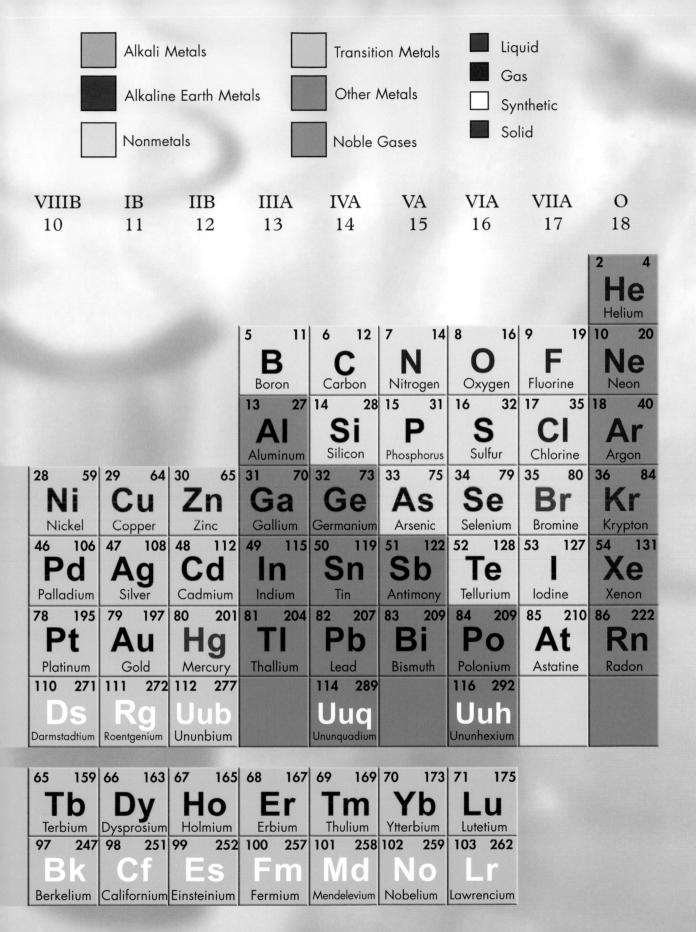

Glossary

carcinogen A substance that can cause cancer.

catalyst A substance that increases the rate of a chemical reaction without itself changing.

contaminate To make something unclean or polluted.

half-life The time a radioactive substance takes to lose half of its radioactivity through decay.

hormone A chemical produced by the body that affects how an organ or organs function.

hygienic Promoting health and cleanliness.

immerse To put something in a liquid so that the surface of the liquid covers it completely.

impurity A substance that prevents another substance from being homogeneous.

metabolize To cause something to change through chemical reactions that take place within the body.

nasal septum The cartilage that separates the right and left sides of the nose.

pH A scale that expresses how acidic or alkaline a solution is. Short for potential of hydrogen. Values range from less than zero to more than 14.

pigment A substance composed of tiny particles that is added to something to give it a specific color.

radioactive Referring to a substance that emits high-energy radiation, and often particles, as unstable atoms in it decay.

supplement A substance or medicine taken to make up for a lack of a mineral in one's diet.

American Chemical Society
1155 16th Street NW
Washington, DC 20036
(800) 227-5558 or (202) 872-4600
Web site: http://www.chemistry.org/portal/a/c/s/1/home.html
An organization whose Web site is a portal dedicated to all aspects of
  chemistry, including its study and practice.

British Stainless Steel Association (BSSA)
Broomgrove
59 Clarkehouse Road
Sheffield, S10 2LE
United Kingdom
Web site: http://www.bssa.org.uk
An organizations whose members promote and aid in the production of
  stainless steel.

International Chromium Development Association (ICDA)
45 Rue de Lisbonne
75008 Paris
France
Web site: http://www.icdachromium.com
The ICDA is an organization dedicated to the sustainability of chromium.

Stainless Steel Information Center (SSIC)
Specialty Steel Industry of North America
3050 K Street North
Washington, DC 20007
(202) 342-8630
Web site: http://www.ssina.com
An organization devoted to the production of and education about
    stainless steel.

Web Sites

Due to the changing nature of Internet links, Rosen Publishing has
developed an online list of Web sites related to the subject of this book.
This site is updated regularly. Please use this link to access the list:

http://www.rosenlinks.com/uept/chro

For Further Reading

Newmark, Ann, and Laura Buller. *Chemistry.* New York, NY: DK Children, 2005.

Parker, Steve. *Rocks and Minerals.* New York, NY: DK Children, 1997.

Pough, Frederick H. *Peterson First Guide to Rocks and Minerals.* New York, NY: Houghton Mifflin, 1991.

Stwertka, Albert. *A Guide to the Elements.* New York, NY: Oxford University Press, 2002.

Wiker, Benjamin D. *The Mystery of the Periodic Table.* Bathgate, ND: Bethlehem Books, 2003.

Bibliography

Brockovich, Erin. "Erin Brockovich Biography." Retrieved June 20, 2007 (http://www.brockovich.com/bio.htm).

Chromium Picolinate. "Dietary Needs." Retrieved June 12, 2007 (http://www.chromiumpicolinate.org/DIETARY_chromium_rich_foods.htm).

Committee on High-Purity Electrolytic Chromium Metal. *High-Purity Chromium Metal: Supply Issues for Gas-Turbine Superalloys.* Washington, DC: National Academy Press, 1995.

Emsley, John. *Nature's Building Blocks: An A–Z Guide to the Elements.* New York, NY: Oxford University Press, 2001.

Glasser, Jeff. "Chrysler Building." Retrieved June 6, 2007 (http://www.nyc-architecture.com/MID/MID021.htm).

Health Encyclopedia: Diseases and Conditions. "Chromium." Retrieved June 13, 2007 (http://www.healthcentral.com/encyclopedia/408/402.html).

Hess, Christopher, and Paul Lehnert. "Chromium." July 7, 2005 Retrieved June 13, 2007 (http://health.yahoo.com/ency/healthwise/ut1024spec).

International Chromium Development Association. "Chromium and Health: A Summary." Retrieved June 19, 2007 (http://www.icdachromium.com/pdf/chromium/Chromium&Health.pdf).

International Chromium Development Association. "Historical Background." Retrieved June 12, 2007 (http://www.icdachromium.com/chromium-historical.php).

International Chromium Development Association. "A Unique Ingredient." Retrieved June 12, 2007 (http://www.icdachromium.com/chromium-introduction.php).

Jones, Chris J. *d- and f-Block Chemistry*. New York, NY: John Wiley & Sons, 2002.

Knapp, Brian. *Iron, Chromium, and Magnesium*. Danbury, CT: Grolier, 1997.

Lepora, Nathan. *Chromium*. New York, NY: Marshall Cavendish, 2006.

Mineral Information Institute. "Chromium." Retrieved June 12, 2007 (http://www.mii.org/Minerals/photochrom.html).

Stainless Steel Information Center. "Stainless Steel Overview: Features & Benefits." Retrieved June 12, 2007 (http://www.ssina.com/overview/features.html).

Stainless Steel Information Center. "Stainless Steel Overview: History." Retrieved June 12, 2007 (http://www.ssina.com/overview/history.html).

Zimmerman, Emily. "Building the Chrysler Building: The Social Construction of the Skyscraper." American Studies at the University of Virginia. Retrieved June 6, 2007 (http://xroads.virginia.edu/~1930s/DISPLAY/chrysler/Frame-1.html).

Index

About the Author

Greg Roza is a writer and editor who specializes in creating library books and educational materials. He lives in Hamburg, New York, with his wife, Abigail; his son, Lincoln; and his daughters, Autumn and Daisy. He has a master's degree in English from SUNY Fredonia. He likes to read about science and do science experiments with his kids. Roza has written several books for Rosen Publishing, including others about the periodic table.

Photo Credits

Cover, pp. 1, 14, 16, 18, 40–41 by Tahara Anderson; p. 5 © AFP/Getty Images; p. 8 © Roger Viollet/Getty Images; p. 11 © Mauro Fermariello/Photo Researchers, Inc.; p. 22 (left) © Biophoto Associates/Photo Researchers, Inc.; p. 22 (right) © Charles D. Winters/Photo Researchers, Inc.; p. 24 © TASS/Sovoto; p. 26 © SSPL/The Image Works; p. 29 Wikimedia Commons; p. 30 © Dirk Wiersma/Photo Researchers, Inc.; p. 33 © Getty Images; p. 35 © Car Culture/Corbis; p. 36 © T. Banner/Custom Medical Stock Photo; p. 37 Shutterstock.com; p. 38 © AP Images.

Designer: Tahara Anderson; **Editor:** Peter Herman